JAN 1 3 2021

Middleton Public Library
7425 Hubbard Avenue
Middleton, WI 53562

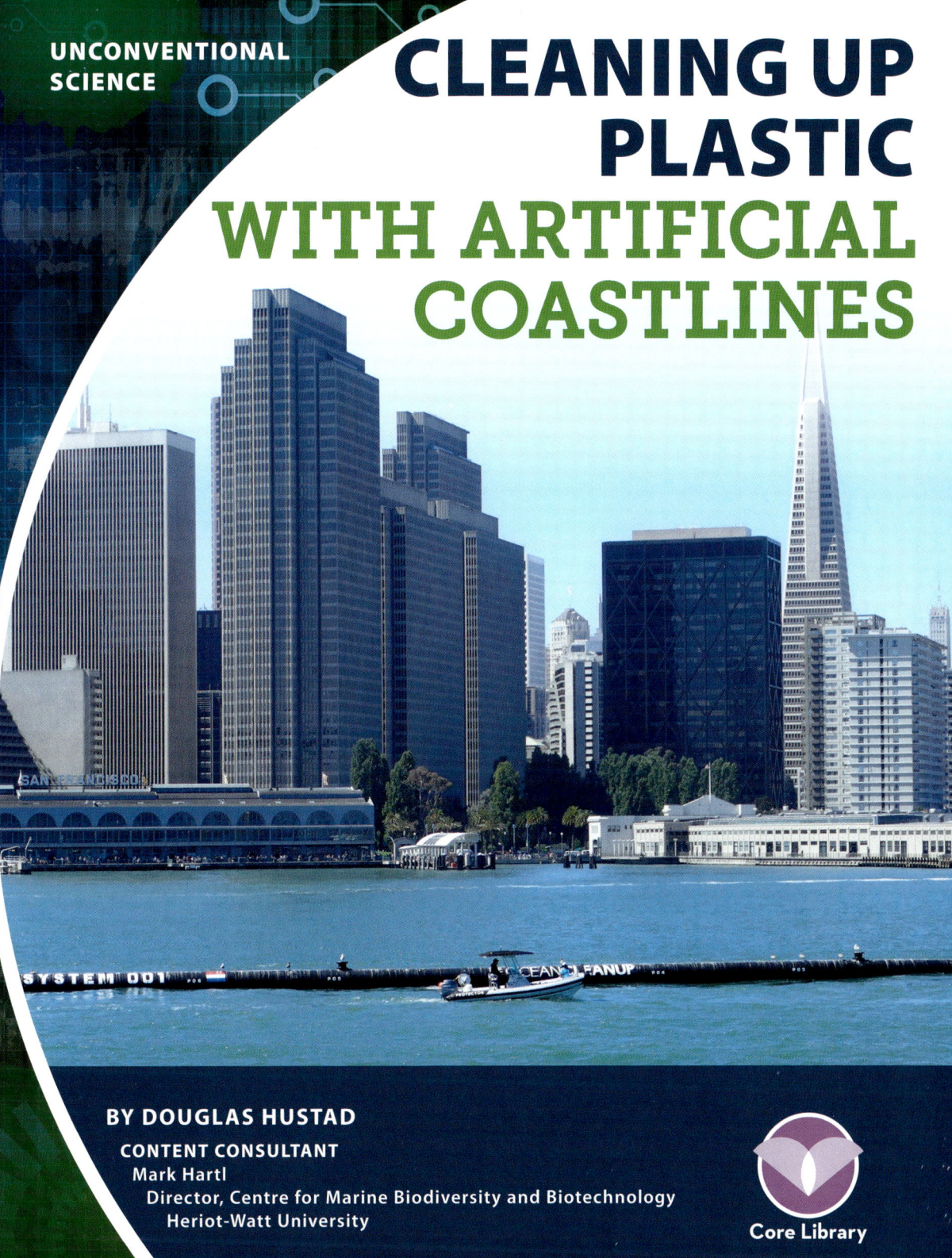

UNCONVENTIONAL SCIENCE

CLEANING UP PLASTIC WITH ARTIFICIAL COASTLINES

BY DOUGLAS HUSTAD

CONTENT CONSULTANT
Mark Hartl
Director, Centre for Marine Biodiversity and Biotechnology
Heriot-Watt University

Cover image: The Ocean Cleanup system launched on September 8, 2018.

Core Library
An Imprint of Abdo Publishing
abdobooks.com

abdocorelibrary.com

Published by Abdo Publishing, a division of ABDO, PO Box 398166, Minneapolis, Minnesota 55439. Copyright © 2020 by Abdo Consulting Group, Inc. International copyrights reserved in all countries. No part of this book may be reproduced in any form without written permission from the publisher. Core Library™ is a trademark and logo of Abdo Publishing.

Printed in the United States of America, North Mankato, Minnesota
042019
092019

Cover Photo: Barbara Munker/picture-alliance/dpa/AP Images
Interior Photos: Barbara Munker/picture-alliance/dpa/AP Images, 1; iStockphoto, 4–5, 40; The Ocean Cleanup, 7, 26, 30–31, 33, 36–37, 45; Carl Nesensohn/AP Images, 9; Jonathan Alcorn/Bloomberg/Getty Images, 12, 43; Peter Bennett/Science Source, 14–15; Red Line Editorial, 16; Rich Carey/Shutterstock Images, 18; Annette Birschel/picture-alliance/dpa/AP Images, 20; Remko de Waal/AFP/Getty Images, 22–23; Josh Edelson/AFP/Getty Images, 28

Editor: Marie Pearson
Series Designer: Ryan Gale

Library of Congress Control Number: 2018965964

Publisher's Cataloging-in-Publication Data

Names: Hustad, Douglas, author.
Title: Cleaning up plastic with artificial coastlines / by Douglas Hustad
Description: Minneapolis, Minnesota : Abdo Publishing, 2020 | Series: Unconventional science | Includes online resources and index.
Identifiers: ISBN 9781532118975 (lib. bdg.) | ISBN 9781532173158 (ebook) | ISBN 9781644940884 (pbk.)
Subjects: LCSH: Pollution control equipment--Juvenile literature. | Recovery of waste materials--Juvenile literature. | Pollution--Prevention and control--Juvenile literature. | Environmental protection--Juvenile literature. | Protection of nature--Juvenile literature.
Classification: DDC 363.7382--dc23

CONTENTS

CHAPTER ONE
A Shore Offshore 4

CHAPTER TWO
The Discovery 14

CHAPTER THREE
How It Works 22

CHAPTER FOUR
The Cleanup Begins 30

CHAPTER FIVE
What's Next? 36

Fast Facts 42

Stop and Think 44

Glossary 46

Online Resources 47

Learn More 47

Index 48

About the Author 48

CHAPTER ONE

A SHORE OFFSHORE

With every lap of the waves, more plastic gathers. Some of it is in the form of tiny particles. Some is larger, such as a milk jug. Some is larger still, such as a fishing net.

Currents push this plastic junk around. The plastic is distributed throughout the ocean. But it collects in large amounts in several spots in the middle of Earth's oceans. One spot is the Great Pacific Garbage Patch. Larger than the state of Texas, the patch contains trillions of pieces of plastic. It is located between Hawaii and California and is growing in size every day. Boats can pass through parts of it

Plastic pollution is a big problem in bodies of water around the world.

HISTORY IN PLASTIC

The Great Pacific Garbage Patch contains approximately 87,000 short tons (79,000 metric tons) of plastic. This material comes from all over the world. A 2018 study found some of it has been there for decades. Items found had 12 different countries of origin. Nearly half of the material is abandoned equipment from commercial fishing boats. Some came from Japan in the wake of the 2011 earthquake and tsunami.

without even knowing it is there. Much of the plastic consists of small particles suspended just below the surface. Nowhere is the plastic nearly dense enough to stand on. And it cannot be seen from space.

THE OCEAN CLEANUP

Nearly 9 million tons of plastic ends up in Earth's oceans each year. The Ocean Cleanup aims to help reverse the damage. A series of booms, each 0.6 miles (1 km) long, floats with the ocean current. Booms are long, floating barriers. These booms act as an artificial coastline,

The Ocean Cleanup's system may be part of the solution for a cleaner ocean.

collecting plastic pieces of all shapes and sizes. Those underwater get caught in screens that hang from the floating booms.

The plastic then gets funneled to a collection center. From there, ships take it to land. They bring it to a facility to process the plastic. Some of it will become raw materials to make new products. Some of it will be used to make fuels. Day and night, slowly the ocean will hopefully get a little bit cleaner.

For now, an ocean free of plastic is science fiction. If everything goes according to plan, the Great Pacific Garbage Patch won't even be half cleaned up until 2024. And that is just one patch among many on Earth. It will not be easy or fast. But if it works, it will help reduce marine plastic pollution.

A HISTORY OF PLASTIC

Alexander Parkes invented the first plastic material in 1885. But it wasn't until World War II (1939–1945) that plastic saw widespread use. With so much

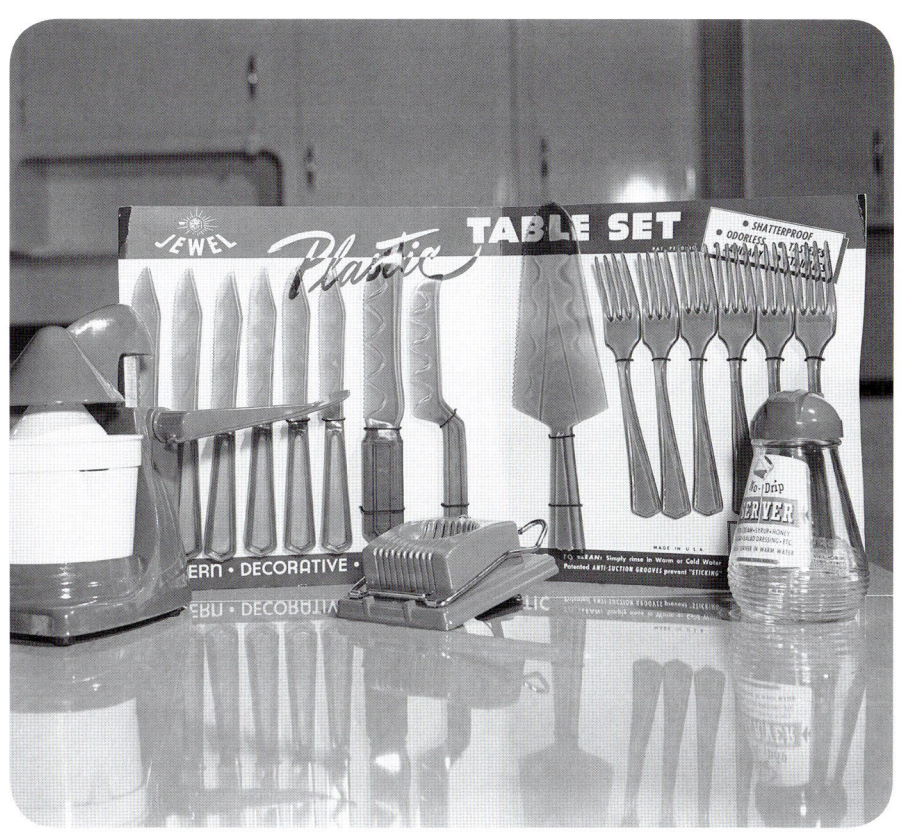

Plastic silverware quickly became popular during World War II.

manufacturing needed for the war, plastic was a cheap alternative. After the war, the plastic factories shifted to making new products.

Plastic was seen as disposable. It was easy to make a container out of plastic that would get thrown away after one use. This increased the amount of trash that people made. Since the 1950s, there have been billions

HENDERSON ISLAND

The consequences of plastic pollution are felt even in places where there are no humans. Henderson Island is a remote island in the South Pacific. Nobody lives there. Yet 3,500 pieces of plastic trash wash up on its shores each day. Forty million pieces have washed up in total. That amounts to 18 short tons (16 metric tons) of plastic. That is the highest density recorded on Earth.

of tons of plastic produced. And almost all of it gets thrown away. Less than 10 percent is recycled. When plastic is thrown away, it usually goes to a landfill.

But plastic can also blow away while being transported to a landfill. Some people litter, and plastic enters the sewer system. From there, it may end up in lakes and rivers. These eventually feed to the ocean. Additionally, huge container ships cross the oceans every day. Sometimes they lose products into the sea. Fishing boats also lose gear, or fishers illegally throw it away into the water.

MICROPLASTIC

Some plastic gets in the ocean even more easily. Microplastics are small pieces of plastic 0.2 inches (5 mm) long or smaller. There is so much of it that a lot escapes water filter systems at treatment plants. The pieces pass back into waterways. Microplastics come from a few sources. Some are microbeads that are added to personal care products such as face scrubs and toothpaste. A lot is from larger plastics breaking apart over time. But some small pieces come from the plastic-making process. Most plastics are made from fossil fuels and shaped into small pellets. These pellets are melted down to make bottles and other products. Sometimes the pellets never reach production. They may spill from shipping containers. Many end up in the ocean.

Microplastics are a big part of the ocean plastic problem. There are several hundred thousand tons of them in the ocean. It is harder to clean up these

A sample from the Great Pacific Garbage Patch shows the high concentration of microplastics in the water.

particles than the larger debris that floats near the surface.

Microplastics are especially dangerous for marine life. Fish and other animals can't see the plastic.

The pieces can begin to smell like food after being in the water for a while. Fish and seabirds accidentally or intentionally eat these plastics. When animals eat plastic, it can make them sick or kill them. And when larger animals and even people eat fish that have consumed plastic, they may be exposed to toxic chemicals in the plastic.

FURTHER EVIDENCE

This chapter defines the Great Pacific Garbage Patch and the problem of plastic pollution. Review the chapter and identify some of the patch's characteristics. Then go to the link below and view the video or read the transcript. Does the video support the information in the chapter? Was there any new information?

WHAT IS THE GREAT PACIFIC GARBAGE PATCH?
abdocorelibrary.com/artificial-coastlines

CHAPTER
TWO

THE DISCOVERY

Once plastic gets into the ocean, currents determine where it goes. Ocean currents lead to one of five different collection points called gyres. A gyre is a big swirling whirlpool of current.

The Great Pacific Garbage Patch is located in the North Pacific Gyre. The four other gyres are located in the North Atlantic, South Atlantic, South Pacific, and the Indian Ocean. These gyres help move ocean waters around Earth.

The Great Pacific Garbage Patch is the gyre of biggest concern. But since 2013, the South Pacific and North Atlantic gyres have seen increasing amounts of marine debris.

Plastic waste big and small collects in ocean gyres around the world.

THE WORLD'S GYRES

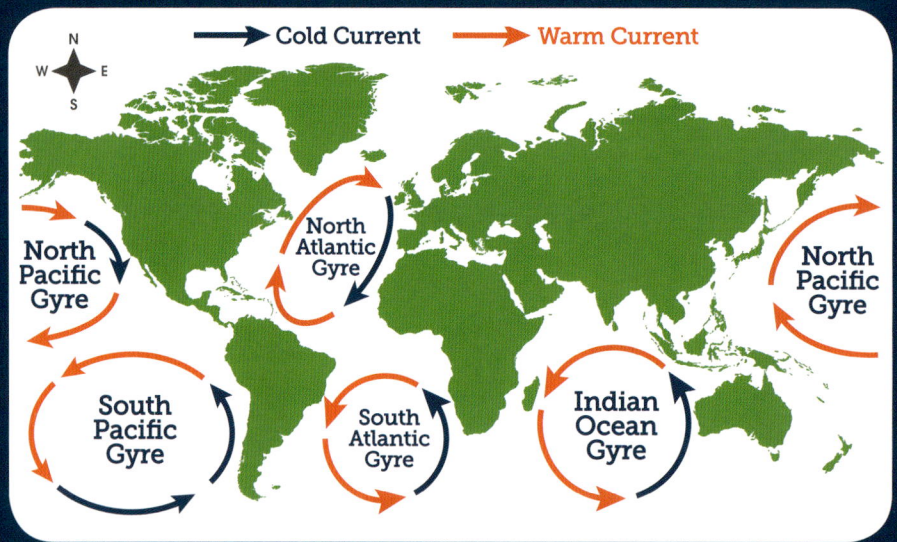

The Great Pacific Garbage Patch is the biggest and most well-known collection point for ocean plastic. But there are four other major ocean gyres where plastic collects. Take a look at this map of the five major ocean gyres. How do you think cleaning them up will help the environment?

These collection points are one place to focus cleanup. The world's oceans are huge. But cleaning up these gyres will get rid of some of the trash.

AN IDEA

It is one thing to read about plastic pollution. By 2011, it was a well-known problem. But 16-year-old Boyan Slat saw the problem for himself that year. Slat was scuba diving in Greece. He saw plastic bags floating in the water. There were more of them than there were fish. It was alarming. Slat had one simple thought: "Why don't we just clean it up?"

Slat went back to his high school in the Netherlands. Cleaning up the ocean became his science project. Over the next six months, he studied the problem. He looked at what had been done before. Most of the proposals for ocean cleanup focused on manually collecting the plastic. This means using a fleet of ships and collection devices

BAGS EVERYWHERE

Floating plastic bags inspired Boyan Slat to clean up the ocean. Plastic bags are one of the most common uses of plastic. People around the world use 500 billion of them each year. People use one million each minute. Some cities have banned plastic bags from stores to reduce waste. This reduction in plastics is another way to tackle the problem of plastic pollution. San Francisco, California, was the first to do so in 2007. By 2018, more than 300 cities had banned the bags or charged a tax to use them.

Plastic can kill coral and ocean wildlife.

to pick it all up. Slat saw that there were a few problems with this.

First, it would be expensive. Gyres like the one in the North Pacific are thousands of miles offshore. Many ships would have to travel a long way to go back and forth. They would all need crews, fuel, and supplies. The ships would give off emissions that are bad for the environment. It also would be very slow. To clean it all, it would take thousands of years.

It also would miss a lot of plastic. Cleaning up the ocean was not as simple as scooping up milk bottles with a fishing net. A lot of the microplastics floated below the surface. The plastic also moves with the current. It does not sit in one spot waiting to be collected.

All these issues gave Slat an idea. Why go to the plastic when the plastic will come to you? Lots of plastic washes up on beaches. Slat designed a system that would use the ocean current to collect the plastic. The plastic would be disposed of later.

TELLING THE WORLD

The idea was simple. The thinking behind it made sense. In October 2012, Slat

OTHER WATERWAYS

The Ocean Cleanup's technology is made to work in the ocean. This is where most plastic ends up. It is a huge problem, but there is also a big problem with pollution in rivers and lakes. They can contain just as much plastic per unit of water as the ocean. They are the main source of ocean plastic. This can cause toxic chemicals to enter local water supplies.

Boyan Slat researched plastic pollution when planning his system.

gave a presentation at a science conference. At first, it did not get much attention.

But Slat still believed in the idea. He talked with experts to see if they thought it could work. With only a few hundred dollars in savings, Slat started an organization called the Ocean Cleanup. It was dedicated to bringing his idea to life.

Then in March 2013, Slat's presentation got noticed. A video of it was published on several news sites. People around the world became interested. They wanted to be a part of it. The Ocean Cleanup was able to hire a team of people. It raised $90,000 to start its initial research, all through online donations. The Ocean Cleanup had taken its first step out to sea.

EXPLORE ONLINE

Chapter Two explains Boyan Slat's idea about how to clean up the ocean. He did a lot of research to see what experts thought. How was Slat's idea different from other proposals? Check out the website below for another organization dedicated to cleaning up the ocean. How are they different from the Ocean Cleanup? What methods do they use?

5 GYRES: SCIENCE TO SOLUTIONS
abdocorelibrary.com/artificial-coastlines

CHAPTER
THREE

HOW IT WORKS

Slat wanted to create a system of long, floating booms anchored to the ocean floor. The booms would form a U shape to gather plastic in the center. Nets hanging below the booms would ensure the plastic wouldn't escape underneath. The booms also had to be thick enough so no plastic would float over them.

Once collected, ships would come and take the plastic to shore. There it would be recycled for another use. But even such a simple design would take a lot of time and money to create.

The booms on the early test models of the system were a bit different from those of the final system.

RESEARCH AND TESTING

Slat was grateful for the interest in his idea in 2013. But with that interest came a lot of criticism and questions. People were not sure if his idea would actually work. So he spent some time trying to find out.

In 2014, Slat recruited a team of nearly 100 scientists and experts to help. They found that the Ocean Cleanup was possible. But it would cost approximately $40 million per year to operate. The company raised more than $2 million from online donations by September 2014. The money allowed more research.

> **MOVING GYRES**
>
> Most gyres on Earth stay in about the same spot. But some move with the seasons. The Indian Ocean Gyre's current changes direction. In the summer, when winds blow north from the Indian Ocean, it moves clockwise. In the winter, winds blow south from the mountains of Tibet. The current moves counterclockwise.

DOING RESEARCH

In 2015, the Ocean Cleanup launched several trips to study plastic in the oceans. The company measured plastic at various depths. This helped the company determine how long the screens needed to be.

A much bigger trip took place the same year. The Ocean Cleanup took 30 ships across the Great Pacific Garbage Patch. They mapped the entire 1.4-million-square-mile (3.5-million-sq-km) area to see what they were dealing with.

TESTING STAGE

The Ocean Cleanup needed to test the system in the ocean. In June 2016, the Ocean Cleanup launched a 330-foot (100-m) test

THE VAST OCEAN

One of the biggest problems with cleaning up the ocean is its size. The oceans are huge. They total 140 million square miles (361 million sq km). That is 70 percent of Earth's surface. And that is just the surface. The oceans are an average of 2.3 miles (3.7 km) deep. Humans have explored only 5 percent of the ocean.

The Ocean Cleanup flights used technology to help find plastic. This included a camera, *circle on the right*, that detected light waves the human eye can't see.

boom in the North Sea. The North Sea was a good testing place. The waters were rougher than in the Pacific. If it could survive the North Sea, it could do well in the North Pacific Gyre. To make the boom easier to observe, it was placed only 14 miles (23 km) offshore. The real thing would be thousands of miles offshore.

By August, the floating booms had taken some damage. The team ended the test early to improve the system. They changed to a tougher material and also went to a free-floating design.

Later that year, the team took some flights over the Pacific. It mapped the extent of the plastic from the air for the first time. They learned where the biggest pieces were and what areas to focus on.

READY TO LAUNCH

By 2018, a second real-world test was ready. Now based in San Francisco, California, the Ocean Cleanup sent a test section approximately 57 miles (92 km) off the California coast. The test section was nearly 400 feet (120 m) long. This test was also the first with a screen attached.

The first step in the test was to see how well a ship could tow the boom. Then the company would take it back to land and finish building the entire boom. A ship would then bring it farther offshore as a full-scale test. If all went well, the rest of the boom would be deployed later that year. The tow test was a success. By June, the rest of the system was being built in San Francisco.

Two weeks of testing took place approximately 400 miles (644 km) off the California coast in September 2018. The test proved the U shape was good at

A ship towed the system under the Golden Gate Bridge and out to sea.

collecting plastic, the system moved quickly in the water, and it could adjust course based on the wind and waves. The net got a couple of tears. These likely happened during transit. But they did not seem to affect the system's ability.

After that successful test, the team decided the system, now known as System 001, was ready. In early October, the team began towing System 001 to the heart of the Great Pacific Garbage Patch.

STRAIGHT TO THE SOURCE

Slat was well aware of the challenges of cleaning up the ocean. He acknowledged that his system was just one of the things that had to be done. But he said it was still a vital part:

> *We also need to prevent more plastic from entering the oceans in the first place, but, that is not a substitute to cleaning up plastic from the deep sea because, once in those offshore accumulation zones, the plastic doesn't really go away by itself. With our expeditions, we have found plastics, really, dating back to the 1970s even. . . . It also gets more harmful over time.*
>
> Source: "Ambitious Effort to Clean Plastic from the Ocean to Start Next Year." *WBUR*. WBUR, May 30, 2017. Web. Accessed September 20, 2018.

What's the Big Idea?
Read this primary source carefully. What is the main point he is trying to make? How does he support that point?

CHAPTER FOUR

THE CLEANUP BEGINS

The Ocean Cleanup learned a lot in the research and testing phase. The biggest change to the system was making it float freely in the water. Floating with the current has a few advantages. The system is able to absorb the force of waves better. That means less damage and fewer repairs.

It works better too. By floating like the plastic, it is able to capture more plastic. Waves, wind, and ocean currents move the system. Plastic is only moved by the current, so it moves slower. The system uses its speed to catch the plastic and gather it up.

The booms are designed to handle rough waters.

HELPFUL DEBRIS

Not all marine debris is bad for the environment. Sometimes a large piece of debris can sink and form an artificial reef. This can be done intentionally too. Old ships and even subway cars have been sunk for this purpose. Great care must be taken to make sure the debris is cleaned of anything that could be toxic. This should only be done by experts.

BIG BOOMS

The booms themselves are made of plastic. This form of plastic is very durable. But it can still bend with the waves. The booms collect plastic on the water's surface.

Hanging below the booms are the screens. The screens are vital for catching microplastics. Many microplastics float just beneath the surface. But they do not go very deep. At approximately 16 feet (5 m), there are fewer particles. The Ocean Cleanup's screen was set to be approximately 5 feet (1.5 m) deep. That is long enough to catch many microplastic particles.

Because of microplastics, the screens have to have very small openings. It looks almost like a window

SYSTEM DIAGRAM

The Ocean Cleanup system uses technology to help it function and prevent ships from running into it. Why do you think these pieces of technology are important? What might happen if one was removed?

- Satellite Pods: allow the Ocean Cleanup to track the system's location

- Navigation Pods: alert nearby ships of the system's location

- Camera: allows the Ocean Cleanup to remotely rotate the view 360 degrees to see footage of the current conditions in the area

- Lanterns: provide light so the system can be easily seen; they also contain technology that allows ships to detect them

screen. Most sea life is too large to pass through it. But some can swim under the screen. There is a weight on the bottom to keep the screen straight up and down.

SMART SYSTEM

Ships tow the system out to sea to get it started. Once at sea, the system reports information back to shore. The Ocean Cleanup team can see where it is and where it's going. Then the company can adjust it if things go off course.

Since the system moves with the ocean, it doesn't need any kind of motor. The only power the system needs is for lights and for the monitoring equipment. All that energy comes from solar power.

The system sends a signal when it is full. The Ocean Cleanup team then sends out a ship to pick it all up. Once the plastic is back on land, some of it will be recycled.

SYSTEM ISSUES

The system had a rough start. After four weeks in the ocean, researchers noticed that plastic would enter and then exit the system. Researchers thought the U shape was too tight. They opened up the arc more. They hoped this would keep plastic from bouncing out. But it didn't work. They tried to find another solution. But before they were able to fix the problem, another one formed. One of the system's ends broke off on December 29, 2018. They towed the system back to shore for repairs. They tried to make the system work like they envisioned. Researchers hoped they could learn from System 001. They wanted to launch a second system in 2020. This system would improve on the first one.

ALERTING SHIPS

The Ocean Cleanup system lets ships know where it is. It has lights and also shows up on radar. Warnings get broadcast over radio to let ships know the Ocean Cleanup is operating nearby.

CHAPTER
FIVE

WHAT'S NEXT?

The Ocean Cleanup hopes to eliminate half the plastic in the Great Pacific Garbage Patch in five years. If that goes well, the system will be rolled out around the world. If people also start using less plastic and creating less new pollution, the ocean could have much less plastic by 2050.

From the start, people questioned Slat's idea. Those who question it are unsure whether or not a teenager is really capable of solving a problem so big. The idea is very different than what has been tried before.

Earlier efforts were pretty simple. Ships just made passes through the water, picking up plastic in nets. But, as Slat had discovered,

While some people believe that the system will do more harm than good, others hope that it is a step toward plastic-free oceans.

CUTTING PLASTIC

Experts play a role in reducing plastic. They can show people ways that they can reduce the amount of plastic they use. Companies can also help reduce the amount of plastic produced. Several companies are phasing out plastic straws. Starbucks is one of these companies.

that was a difficult and slow process.

THE CRITICS

Even with the Ocean Cleanup's research, some experts in the field are still skeptical. They think that cleaning up the oceans is unrealistic and almost impossible. Other concerns include damage to wildlife and what to do with the collected plastic.

The Ocean Cleanup is designed for marine life to swim under it. But some species may not be able to do so. They could get trapped. Or they could be caught in machines collecting the plastic.

Also, recyclers may not want all of the collected plastic. Some of the plastic is very old and has started to break up or is the wrong type of plastic. It may not be suitable for recycling. There is no easy way to get rid of unwanted plastic. Burning it can release toxic chemicals into the air.

HOW TO HELP

Experts say individuals are the best way to protect the environment from plastic. The more people reduce the plastic they use, the less plastic will end up in the ocean. It may seem like one person cannot do much. But one person can help a lot.

Shop with reusable bags instead of plastic ones. This saves an average of 307 bags per person each year. If you don't need to, don't use straws. Don't drink from plastic bottles that get thrown away after one use. Use reusable

THE SEABIN

The Seabin is a small-scale project. It looks like a trash can floating in the water. The Seabin bobs up and down, collecting plastic that flows into it. It cannot clean up a whole garbage patch by itself. But the technology could be applied to bigger projects.

One good way to keep plastic from polluting bodies of water is by shopping with reusable bags instead of plastic ones.

bottles instead. This can save an average person from tossing 156 bottles a year. Don't pack a lunch using plastic bags. Use washable containers. And definitely do not litter. Recycle everything that can be recycled. Individuals can make a huge difference when it comes to plastic waste.

There is no reason why multiple methods can't be tried at the same time. The plastic problem is big enough to need many solutions. But Slat is determined to make the Ocean Cleanup work. At the very least, he plans to try.

STRAIGHT TO THE SOURCE

Some people are concerned that the Ocean Cleanup is too complicated and say there are simpler solutions:

> *Skeptics say the [Ocean Cleanup's] idea doesn't make much sense and that collecting trash closer to shore would be more cost effective. "Focusing clean-up at those gyres, in the opinion of most of the scientific community, is a waste of effort," says marine biologist Jan van Franeker of Wageningen Marine Research in the Netherlands. "It's a lot of money to reduce something that disappears in 10 to 20 years, if you stop the input." . . . Critics also worry that the high-tech clean-up project could distract from less glamorous efforts to lessen the use of plastic.*
>
> Source: Erik Stokstad. "Controversial Plastic Trash Collector Begins Maiden Ocean Voyage." *Science.* American Association for the Advancement of Science, September 11, 2018. Web. Accessed September 20, 2018.

Point of View

After reading this quote, read the primary source from Chapter Three. Each person has a different point of view on the Ocean Cleanup. How are they different and why? How are they similar and why? Write a short essay explaining what you find.

FAST FACTS

- Plastic use has been on the rise since World War II. A lot of this plastic gets thrown away after a single use. When plastic doesn't make it into a landfill, it can end up in the ocean. The ocean currents collect plastic in one of five gyres all around Earth. These plastics can be hazardous to marine life and the environment.

- When Boyan Slat was a teenager, he came up with the idea for an artificial coastline to catch plastic. He was diving in Greece and saw plastic bags floating underwater. He designed a system as part of a high school science project and then aimed to put it into production. He called his organization the Ocean Cleanup.

- The system works with a long series of booms. These booms float out in the ocean and collect plastic that washes up. The plastic is then taken to land to be recycled. In October 2018, the system was towed out to the Great Pacific Garbage Patch to begin the first stages of cleanup.

- There are some critics of the Ocean Cleanup. They say that the focus should be on preventing plastic from getting into the ocean in the first place. But Slat has stressed that multiple methods should be used at once.

STOP AND
THINK

Tell the Tale

Chapter Two describes Boyan Slat's inspiration for inventing the Ocean Cleanup. He was diving when he started to wonder about all the plastic floating in the ocean. Write 200 words about a time when you saw something you wanted to change. What did you see? How did it make you feel?

You Are There

Chapter Three describes some of the research that the Ocean Cleanup team did. Imagine you are on a ship with them or in a plane making observations. What would it be like to see all that garbage floating in the ocean? Write a letter home describing what you saw and how you felt.

Say What?

Studying artificial coastlines can mean learning a lot of new vocabulary. Find five words in this book that you had not heard before. Use a dictionary to find out what they mean. Then write the meanings in your own words, and use each word in a new sentence.

Why Do I Care?

It might be easy to ignore the problem of plastic pollution. Even if you live near the ocean, the garbage patches are thousands of miles from the shore. But that does not mean it doesn't affect you. How can ocean garbage patches affect people on land? Why should people help reduce these patches?

GLOSSARY

anchor
to secure firmly in position

artificial
something made by humans that isn't found in nature

current
the movement of water in one direction

debris
broken pieces of something

landfill
an underground space for the collection of garbage

fossil fuel
natural gas and other fuels that have formed underground from the remains of plants and animals

particles
tiny pieces of something

radar
a system to detect other ships or planes in an area

recruit
to ask to become a part of an organization

reef
a long ridge near the surface of the water that provides a habitat for sea life

tsunami
a series of waves produced by an earthquake

ONLINE RESOURCES

To learn more about cleaning up plastic with artificial coastlines, visit our free resource websites below.

Visit **abdocorelibrary.com** or scan this QR code for free Common Core resources for teachers and students, including vetted activities, multimedia, and booklinks, for deeper subject comprehension.

Visit **abdobooklinks.com** or scan this QR code for free additional online weblinks for further learning. These links are routinely monitored and updated to provide the most current information available.

LEARN MORE

Hand, Carol. *Marine Science in the Real World.* Minneapolis, MN: Abdo Publishing, 2017. Print.

Murphy, Julie. *Coral Reefs Matter.* Minneapolis, MN: Abdo Publishing, 2016. Print.

INDEX

booms, 6–9, 23, 26–27, 32

criticism, 24, 38–39, 41

Great Pacific Garbage Patch, 5, 6, 8, 13, 15, 16, 25, 28, 37
gyres, 15–16, 18, 21, 24, 26, 41

Henderson Island, 10

Indian Ocean Gyre, 15, 16, 24
issues, 35

launch, 27–28

microplastics, 11–13, 19, 32

North Pacific Gyre, 15, 16, 18, 26
North Sea, 26

Parkes, Alexander, 8
plastic bags, 17, 39–40

recycling, 10, 23, 34, 38–39, 40
reducing plastic, 17, 38, 39–40
research, 21, 24–25, 35

San Francisco, California, 17, 27
screens, 8, 25, 27, 32–34
Seabin, 39
shape, 23, 27, 35
Slat, Boyan, 16–21, 24, 29, 37, 40
System 001, 28, 35

technology, 19, 33, 34, 35
testing, 25–28

World War II, 8

About the Author

Douglas Hustad is the author of several books on science and technology for young people. Originally from the San Francisco Bay Area, he now lives in North County San Diego with his wife and their two dogs.